Edison Twp. Free Public Library
340 Plainfield Ave.
Edison, New Jersey 08817

My First Science Words

SPACE WORDS

A CRABTREE SEEDLINGS BOOK

Taylor Farley

CRABTREE
PUBLISHING COMPANY
WWW.CRABTREEBOOKS.COM

Earth
(URTH)

Sun
(SUHN)

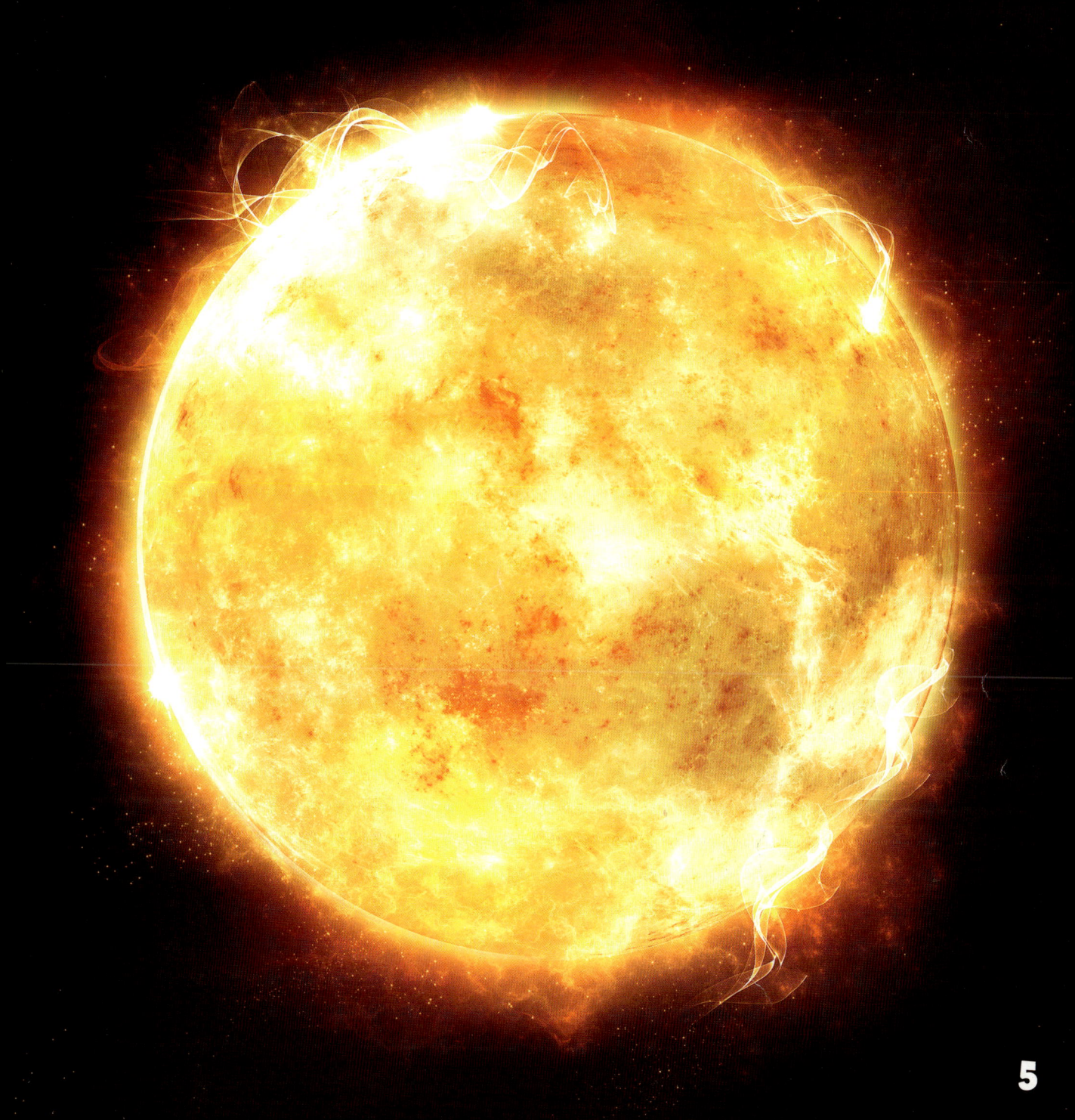

Moon
(MOON)

stars
(STARZ)

planets
(PLAN-its)

telescope
(TEL-uh-skope)

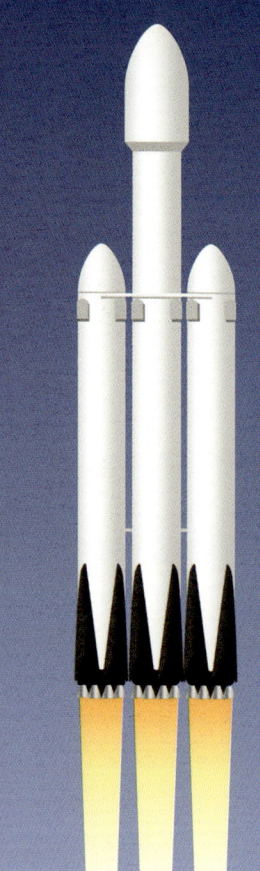

rocket
(ROK-it)

astronaut
(ASS-truh-nawt)

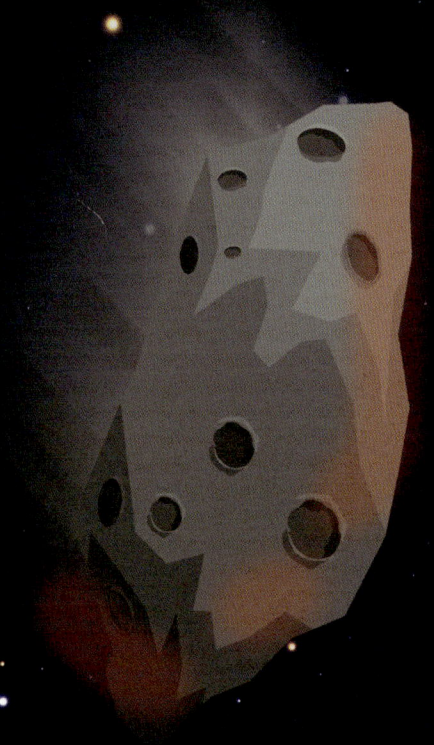

meteor
(MEE-tee-ur)

galaxy
(GAL-uhk-see)

Glossary

astronaut (ASS-truh-nawt): An astronaut is someone who travels in space.

Earth (URTH): Earth is the planet on which we live.

galaxy (GAL-uhk-see): A galaxy is a large group of stars and planets.

meteor (MEE-tee-ur): A meteor is a piece of rock or metal that enters Earth's atmosphere.

Moon (MOON): The Moon is the space object that moves around Earth.

planets (PLAN-its): Planets are large, round objects that circle our Sun. There are eight planets.

Long and straight, from left to right …

The farmer ploughs, day and night.

Wavy lines curve up and down ...

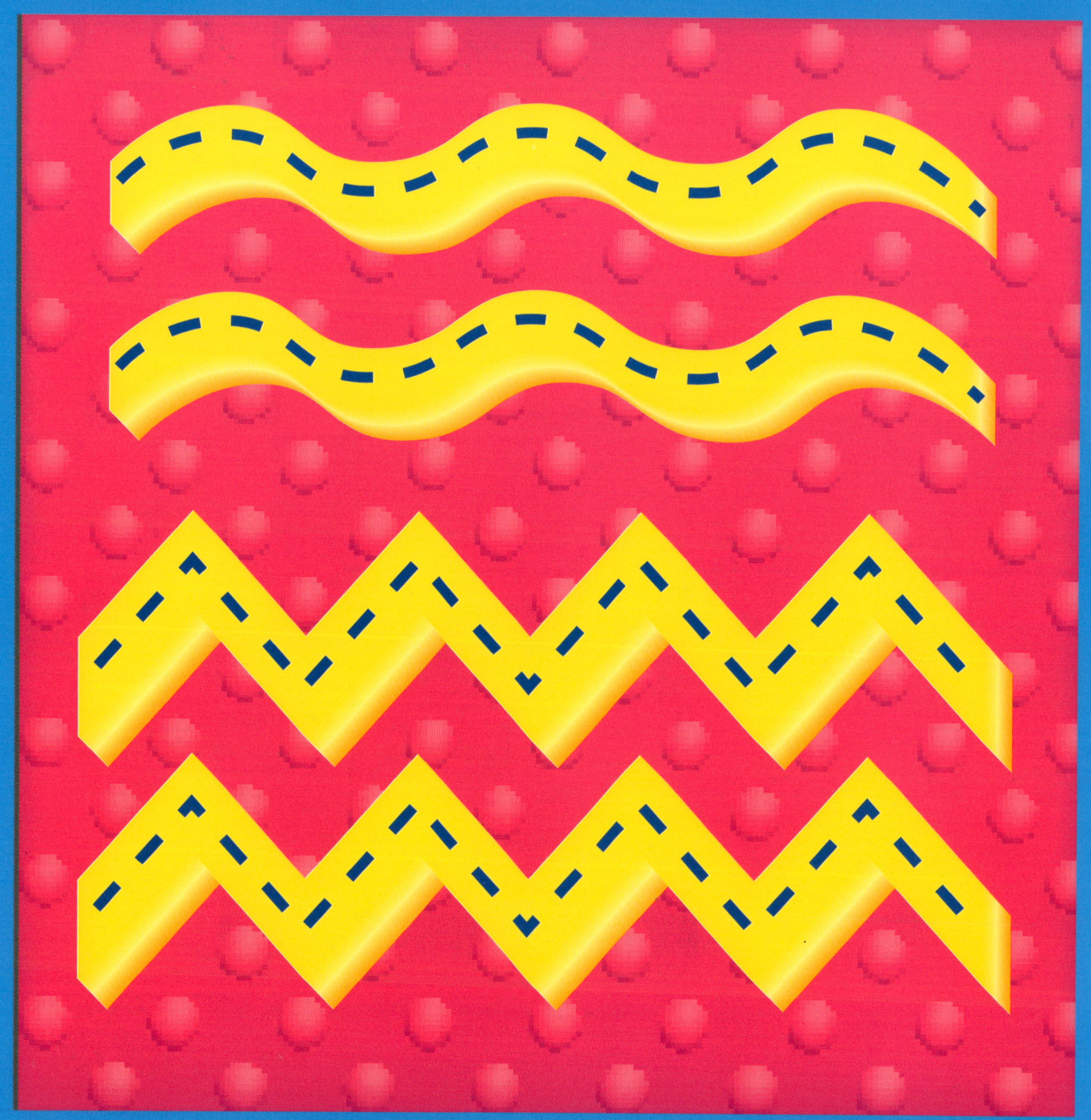

Zig zag, zig zag, through the town.

Bumpy hills that touch the sky ...

Round balloons that fly so high.

Up and over, draw a loop ...

Big trucks flatten, dump and scoop.

Waves rise up and then curl round ...

Steaming, sailing, homeward bound!

Round and round to loop the loop ...